TIDE POOLS

TIDE
POOLS

Carmen Bredeson

A First Book

FRANKLIN WATTS A DIVISION OF GROLIER PUBLISHING

New York   London   Hong Kong   Sydney   Danbury, Connecticut

Expert consultant: Janann V. Jenner, Ph.D.

Photographs©: Photographs ©: Chris Huss: 46, 49; ENP Images: 13, 24 (Gary Braasch), cover, backcover, 1, 3, 7, (Brandon D. Cole), 21 (Gerry Ellis), 8 (D.C. Lowe); Norbert Wu Photography: 25, 28 (Peter Parks); Photo Researchers: 34 (Gary G. Gibson); Tony Stone Images: 27 (Peter J. Bryant/BPS), 18 (Patrick Cocklin), 15 (Darrell Gulin), 2 (Dave Schiefelbein), 36 (Kevin & Cat Sweeney), 54 (Stuart Westmorland), 20; Visuals Unlimited: 52 (A.D. Coplay), 14, 29, 37, 41 (John D. Cunningham), 33 (Robert Degoursey), 11 (Dave B. Fleetham), 30 (Bill Kamin), 43 (Alex Kerstitch), 48 bottom (Ken Lucas), 10, 48 top (Glenn Oliver), 38 (Gustav W. Verderber).

Visit Franklin Watts on the Internet at:
http://publishing.grolier.com

Library of Congress Cataloging-in-Publication Data

Bredeson, Carmen.
     Tide Pools / Carmen Bredeson.
       p. cm. (A First Book)
     Includes bibliographical references and index.
     Summary: Describes the physical characteristics of tide pools and the organisms that inhabit them.
     ISBN 0-531-20368-9 (lib. bdg.)   0-531-15958-2 (pbk.)
     1. Tide pool ecology—Juvenile literature. 2. Tide pools—Juvenile literature. [1. Tide pool ecology. 2. Ecology. 3. Tide pools.] I. Title. II. Series.
QH541.5.S25B724   1999
577.69'9—dc21   97-41630
                                            CIP
                                            AC

CONTENTS

TIDE POOLS

INTRODUCTION

*❝ Life originated in the ocean 3 billion years ago,
and the sea today remains the fountain of life.❞*

Jacques Cousteau, underwater explorer

As the tide rises, water is swept into a series of *tide pools* along the shoreline. Inside one of the pools, fronds of seaweed sway in unison with the ocean current. The seaweed serves as food for many tide pool creatures.

Scattered here and there among the seaweed are what appear to be brightly colored flowers. Their "blossoms" may be beautiful, but they are not harmless. They are the

◀ The incoming tide spills into this tide pool in Shore Acres State Park in Oregon.

▲ A sea anemone

stinging tentacles of an animal called a sea anemone. When an unsuspecting creature brushes against one of the "petals," poisonous darts shoot out and sting the victim. After wrapping its meal in sticky thread, the sea anemone uses the tentacles to draw the prey into its large mouth.

Not all sea anemones live on the tide pool's sandy bottom. Some attach themselves to the back of a crab. The crab doesn't mind, though. The sea anemone's stinging darts protect the crab from enemies. The sea anemone benefits from the union too. It gets the food that is left after the crab finishes its meal.

Most tide pool predators avoid sea anemones, but not the nudibranch. This colorful creature is not affected by a sea anemone's poison. When a nudibranch eats a sea anemone, it digests everything except the stinging cells. These are transferred to a special canal and then to the gills on the nudibranch's back. When an enemy grabs a nudibranch, the attacker gets a powerful dose of the borrowed poison.

In a tide pool, the battle for survival is a never-ending struggle. The animals that live in these small bodies of water have to constantly watch out for *predators*. In addition, their lives are drastically affected by the rise and fall of the tides.

▼ This beautiful nudibranch is poisonous. Its poison comes from the sea anemones it eats.

C H A P T E R   1

WHAT IS A TIDE POOL?

" The shore is an ancient world, for as long as there has been an earth and sea there has been this place of meeting of land and water. "

Rachel Carson, marine biologist

If you've ever been to the beach, you probably spent time watching the ocean's waves roll gently ashore. On some shorelines, however, strong winds whip the waves into a frenzy and send them crashing onto rocky cliffs. Salty oceans and seas cover almost 75 percent of Earth's surface. All over the planet—wherever the ocean meets the shore— the tides rise and fall in a continuous cycle.

As high tides wash up onto the land, they carry a wide variety of sea life with them. Clumps of seaweed, small

▲ Waves crashing into a rocky shore

crabs, barnacle-encrusted driftwood, sea stars—these are just a few of the living things that tumble toward the shore along with the waves. Sometimes they are carried back out into the open ocean, but not always. The small creatures tumbling in the waves are often stranded on shore where, without water, they wither and die.

On some sandy shores, incoming waves spill into low spots along the beach. Where coastlines are rocky, the water fills hollows around or between the rocks. Puddles as small as a bucket or as big as a bathtub form where the water pools.

▲ This mudflat, which contains dozens of tide pools, is completely covered with water at high tide.

These tide pools are miniature worlds full of living things. Some last only a short time before the hot sun dries them up. Others, though, are permanent and provide homes for many different kinds of *marine* life.

Life in a tide pool is not like life in the open ocean. In the huge oceans, water conditions change very slowly. In tide pools, conditions change much more rapidly. And the creatures in tide pools must be able to *adapt* to these changes.

As the water in tide pools evaporates, bright green sea anemones and ▶ colorful sea stars are stranded.

As the sun beats down on a small tide pool, the temperature of the seawater increases very quickly. When the water warms up, it holds less *oxygen*. Since nearly all living things need oxygen to survive, this could mean trouble for the creatures living in the tide pool.

The sun's heat also causes seawater in the tide pool to *evaporate*. Some of the water may turn into vapor, but the salt does not. It stays behind in the remaining water. After several hours, the water in the tide pool has changed dramatically. Originally it was cool, had plenty of oxygen, and was a little salty. Now it is warm, has much less oxygen, and is very salty.

Why don't these harsh conditions kill the creatures in the pool? Sometimes they do, but usually another high tide brings cooler water and a fresh supply of oxygen and food to the animals. For several hours, all will be well in the miniature world. Then, as the tide recedes, the drama to survive begins again.

What are the tides and what causes them to rise and fall like clockwork every single day?

C H A P T E R   2

A CLOSER LOOK AT TIDES

"*Animals that live near the shore must do daily combat with the tides.*"

Jacques Cousteau, underwater explorer

Imagine that you are building a sand castle at the beach. The breaking waves are nowhere near you. You have dumped bucket after bucket of sand onto a growing mound. When the pile is large enough you carefully begin to carve walls and towers. The work is slow because the dry sand crumbles easily.

When the castle is nearly finished, you notice that water is beginning to tickle your toes. Gradually, the waves creep closer and closer. Suddenly, a big wave knocks down one of the towers that you worked so hard to build. Will your masterpiece survive?

You grab a shovel and begin to dig a moat around the castle. Sand flies in all directions! Just as you finish digging, a large wave breaks and gallons of water roll across the sand and spill into the soon overflowing moat.

▼ This sand castle is about to be destroyed be the ocean's powerful surf.

You hop around on the beach, nervously waiting to see what will happen. You are sure your castle will be flattened. But then a miracle occurs—the waves begin breaking farther away. The tide has turned and your castle is saved!

While checking the castle for damage, you notice something moving in the moat. Kneeling down in the sand, you peer into the water and see tiny silver fish darting round and round the moat. You also see a small crab clinging to a piece of seaweed that is trapped among a pile of pearly shells. As you watch, a hermit crab lumbers by. The tide has washed these living things into your moat along with the water. The creatures trapped in the small pool will wage a battle for survival. Whether they live or die depends on the action of the tides.

THE RISE AND FALL OF THE TIDES

Along most shorelines, there are two high tides and two low tides every 24 hours and 50 minutes. The tides rise and fall every 6 hours or so. For example, the high tide that nearly destroyed your castle occurred at 1:00 P.M. That's when the tide reached its highest point before it began to *ebb*, or recede again. The tide continued to ebb until a little after 7:00 P.M. By that time your castle was high and dry, well away from the surf.

▲ The Moon's gravity causes the tides.

While you were sleeping, the tide continued its cycle. It grew higher and higher. At about 1:30 A.M., the waves again lapped at your castle. They brought new water and food to the little animals in the moat. When you woke up and looked out of the window at 7:45 A.M., the sand around the castle was dry again because it was low tide. This same pattern of high and low tides is repeated every day, on beaches all over the world. The exact hours change daily, but the amount of time between high and low tides is always the same.

The rise and fall in Earth's oceans is caused mostly by the gravitational pull of the Moon. Even though the

Moon's *gravity* is only one-sixth as strong as Earth's gravity, it is still powerful enough to pull *matter* toward it. As Earth rotates, the water in the oceans nearest the moon is pulled up in a bulge. On the opposite side of Earth, another bulge forms to balance the planet. These two swells, which produce the high tides, follow the moon as it moves around Earth. In between the two bulges are low areas, called troughs. The troughs create the low tides.

The Sun's gravity also has an effect on Earth's tides. But because the Sun is 93 million miles (150 million km) away, it has less influence on the tides than the Moon. When the Sun and Moon are in line with each other, however, their combined pull creates the highest tides of all. These are called *spring tides* because the water seems to

▼ The Sun's gravity also influences the tides.

"spring" up from Earth. When the sun and moon are at right angles to each other, their gravitational pulls oppose one another. The result is lower than average tides, which are called *neap tides*.

The tides bring a bounty of oxygen and food to the creatures in tide pools. Without this twice-a-day influx of water, none of the animals would survive.

C H A P T E R 3

WHAT THE TIDE BRINGS

"A drop from a pond, viewed through a micro-scope, swarms with tiny organisms, some spinning, some crawling, some whizzing across the field of vision like rockets."

David Attenborough, naturalist

Tides cover and uncover a part of the shoreline known as the *intertidal zone*. This area, which may be sandy, rocky, or grassy, is often underwater during high tides. During low tides, it is out of the water and exposed to the sun, wind, and rain. Tide pools are found in the intertidal zone.

OXYGEN TO BREATHE

When cool water rushes into a tide pool, it brings oxygen with it. During the 12 hours between high tides, oxygen

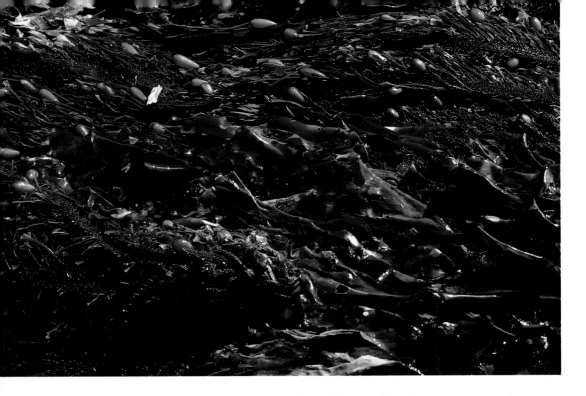

▲ Like plants, seaweed contains chlorophyll and can carry out photosynthesis. Tide pool animals depend on seaweed for oxygen.

levels in the water fall significantly. Some of it is used up by the tide pool creatures when they breathe. In addition, as the sun heats the water in the pool, the water loses oxygen to the air.

During the day, some of the oxygen used by animals is replaced by plants and microorganisms as they make food. In this process, which is called *photosynthesis*, plants and microbes convert water and a chemical called carbon dioxide into a type of sugar known as *glucose*. This process also creates oxygen.

Photosynthesis is powered by energy from sunlight. The energy is absorbed by a substance called *chlorophyll*.

Chlorophyll is found in the leaves and stems of plants and in the cells of some microbes. Because the sun does not shine at night, photosynthesis can only take place during the day. This means that less oxygen is available in a tide pool after the sun sets.

FOOD TO EAT

The incoming tide also brings a new supply of food. The food is primarily in the form of *plankton*, which is made up of millions of tiny creatures. Many are so small that they cannot be seen without the aid of a microscope.

▼ This plankton sample includes young sea stars and sea urchins, fish eggs, and a variety if microorganisms.

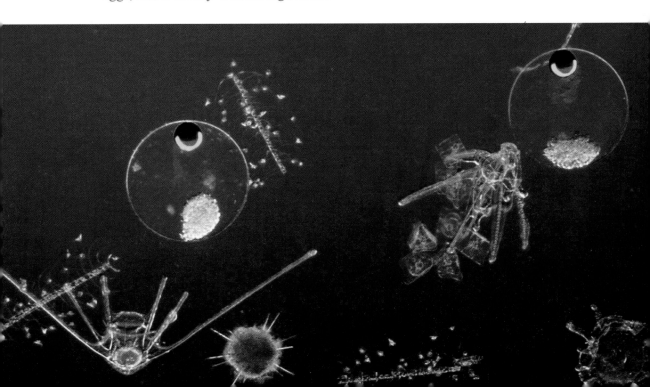

Within the organic mixture, single-celled *amebas* and *diatoms* float among newly hatched shrimp, barnacles, and crabs. Mixed into the plankton are millions of eggs. Since most ocean creatures cannot protect their eggs after spawning, they release huge numbers of them. An oyster, for example, might lay 500 million eggs in 1 year! As many as 99 percent of the eggs deposited in the ocean never reach maturity.

Most *zooplankton* consists of small animals such as water fleas and tiny jellyfish. Zooplankton also includes creatures that are neither plants nor animals, called *protozoans*. These organisms belong to a group of living things called *protists*. Protists are simple creatures that have some characteristics of plants and some characteristics of animals. Most scientists believe that all plants and animals *evolved* from the protists.

Like animals, protozoans—such as amebas—and *ciliates*—such as paramecia—must eat other creatures to stay alive. Some graze on plants, while others devour bacteria or other protozoans.

Phytoplankton consists mostly of simple creatures called *algae*. Like protozoans, algae are protists. All algae have the ability to produce food using photosynthesis. Some also have the ability to move about, and a few can drift or swim.

Diatoms are the most abundant group of algae. They consist of a single cell surrounded by a hard, glassy shell.

▲ Zooplankton often contains water fleas. These tiny creatures are closely related to crabs and lobsters.

They reproduce so quickly that each one may have 1 billion descendants in a month. Their numbers increase so fast because a diatom reproduces by simply splitting in half. Each new diatom has all of the features it needs to live, grow, and reproduce.

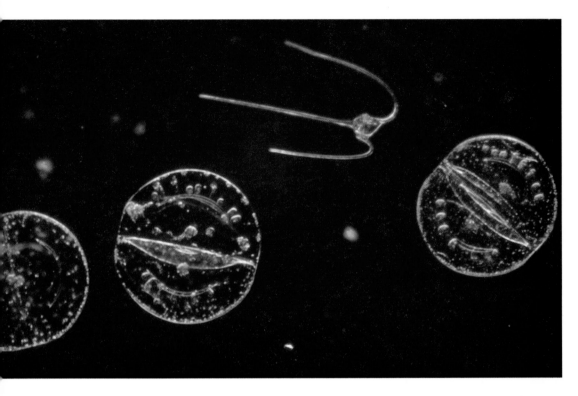

▲ This phytoplankton sample contains a dinoflagellate and several diatoms.

Many types of *dinoflagellates* are also considered algae. Most dinoflagellates have two threadlike whips that allow them to move and spin in the water.

Phytoplankton usually floats near the surface of the water. It is at the base of most food chains in the ocean, just as grass is at the base of many food chains on land. Like grass and other plants, phytoplankton uses photosynthesis to produce all the energy it needs to survive.

Many small marine animals eat phytoplankton. They use the energy stored in the bodies of the algae and other tiny creatures to build the proteins they need to live and grow. In turn, the small ocean animals are eaten by larger animals, including fish. The fish are then eaten by sharks, whales, and other predators.

SEAWEED IS NOT A PLANT

Not all algae are small. If you've ever seen seaweed, you probably thought it was a plant. It isn't. Like other kinds of algae, seaweed is actually a protist. There are two major types of seaweed. Red algae is stringy or feathery. Brown algae, which is also called kelp, looks a lot like lettuce. It grows in dense strands or thick floating mats.

▼ Kelp often attaches itself to rocks along the shore.

Since seaweeds are not plants, they have no flowers, roots, or leaves. Some seaweeds have structures, called *holdfasts*, that look like roots, but do not absorb nutrients. These holdfasts anchor seaweeds in one spot by producing a sticky substance that glues the seaweed to a rock or any other hard surface.

Seaweed is an important part of a tide pool. A clump of seaweed can be a safe home for a thriving community of tiny shrimp, crabs, and sea worms. It gives small creatures a place to lay eggs or hide from enemies, and serves as a source of food for many animals.

◀ Although seaweed does not have roots that absorb nutrients, many types are anchored to surfaces with rootlike structures called holdfasts.

CHAPTER 4

WHAT ARE MOLLUSKS?

"Life in nature is, in essence, a struggle. A struggle for food, for space, for safety, for perpetuation of the species."

Jacques Cousteau, underwater explorer

Many of the creatures that live in tide pools never see their food. They don't have to because they are *filter feeders*. When a wave washes into a tide pool, these animals draw the fresh water into their bodies. They separate or filter the plankton and expel the leftover water, so that more can enter. In this way, filter feeders get all the food they need to live and grow.

These bay scallops are filter feeders. ▶

CLAMS, MUSSELS, OYSTERS, AND SCALLOPS

Clams and mussels are filter feeders, so are scallops and oysters. When a clam senses that new water is entering a tide pool, it quickly pokes two tubes, called *siphons*, out of its hard protective shell. Hairs on the clam's body wave back and forth, directing water into one of the siphons. Inside the clam, the water is filtered so that all the plankton is removed. The wastewater is carried out of the clam through the other siphon. An adult clam may pump 15 quarts (14 L) of water an hour through these tubes.

If feeding is not good in one spot, clams are able to move to another location. By extending a single "foot" from its shell, a clam can pull itself along the seafloor or

burrow into the sand. Even when a clam is buried in sand, it can extend its siphons so that they are above the seafloor. This allows the clam to hide from predators and take in food at the same time.

A clam is one example of a *bivalve*. All bivalves have two shells connected by a strong, muscular tendon. This tendon functions like the hinge on a door. If the animal senses danger, it snaps its two shells shut. Mussels, oysters, and scallops are bivalves, too.

All bivalves are *mollusks*—boneless animals with soft bodies. Snails, slugs, conches, squids, and octopuses are also mollusks. Not all mollusks are found in tide pools, though. Most snails and slugs live on land. Squids and octopuses can swim, so it is unusual—but not impossible—for them to end up in a tide pool.

SQUIDS AND OCTOPUSES

Squids and octopuses belong to a group of mollusks called *cephalopods*. As you probably know, an octopus has eight long arms covered with two rows of suckers. The octopus uses its arms to grab prey and carry the food to its mouth. Most octopuses eat clams, scallops, crabs, and an occasional fish.

◄ A clam uses its long foot to move from place to place.

Sometimes an octopus wiggles the tip of one arm so that it looks like a worm. It then pounces on any unsuspecting creature that tried to get a closer look. An octopus also uses its strong, sharp beaklike mouth to kill small animals. Some kinds of octopuses can inject a poison that paralyzes their victims.

▼ An octopus

When an octopus is attacked by a predator, it draws water into its body cavity and blows it out quickly. The spurt of water shoots the octopus backward, out of harm's way. Often, a cloud of "ink" is also squirted in the direction of the attacker. The inky cloud allows the octopus to escape. In some cases, the ink also numbs the enemy.

A squid has eight arms and two long tentacles growing from its head. Squids often swim in large groups called *shoals*. Like octopuses, squids hide from predators by changing the colors or patterns on their bodies to match their surroundings. A squid can propel itself through the water by filling the folds in its body walls with water and forcing it out through a tube located just below its head.

▼ A squid

MORE MOLLUSKS

Some tide pools contain dogwhelks. A dogwhelk is closely related to a land snail. It has only one shell and crawls around on a footlike structure. Unlike bivalves, a dogwhelk has a distinct head and eyes.

A dogwhelk can be a bivalve's worst nightmare. When a dogwhelk finds a clam, it immediately releases a chemical that softens the shell of its victim. Then, the dogwhelk uses its *radula*—a filelike tongue—to bore a hole through the clam's shell. It may take up to 3 days for the dogwhelk to reach the tasty animal inside.

Not all one-shelled mollusks eat other animals. Limpets, which look like snails wearing pointed hats, are not such mighty hunters. They scrape algae off rocks as they inch along on their muscular "foot." Joining them are periwinkles, which cluster together in large groups on rocks and in cracks. Periwinkles, along with many other mollusks, spend their time gobbling up tiny phytoplankton or munching on seaweed.

◀ Dogwhelks can be quite colorful.

CHAPTER 5

WHAT ARE CRUSTACEANS?

"In every part one meets hermit-crabs of more than one species, carrying on their backs the houses they have stolen . . ."

Charles Darwin, naturalist

Mollusks are not the only animals found in tide pools. They are joined by a group of creatures called *crustaceans,* which often prey on mollusks. These animals have a hard outer shell called an *exoskeleton,* jointed legs, as well as a head with jaws and two antennae. Examples of crustaceans include shrimp, lobsters, and crabs. They breathe by pumping water over gills, just like fish. Within the gills, oxygen is separated from the water and passed into the bloodstream.

▲ Red rock crabs are common along North American coastlines.

The rock crab is a common inhabitant of tide pools. It will eat almost anything, but its favorite meals include worms, seaweed, and dead animals. These crabs are the *scavengers* that keep the tide pool free of rotting matter. To help them find food, rock crabs have two large eyes on the ends of stalks. The stalks also serve as sensitive antennae that help the crabs feel things and smell prey. When a rock crab catches a victim, it uses its powerful claws to tear the prey apart and push the pieces into its strong jaws.

As crabs grow, they eventually get too large for the shell that surrounds their bodies. They must shed the old shell in order for a new one to grow. To begin this process, called *molting*, the crab swallows a lot of water. Its body swells until the top part of the shell separates from the bottom part. The crab then crawls out backward, leaving behind an entire shell—complete with eye stalks. Until its new shell hardens a few days later, the rock crab is an easy target for hungry predators.

Hermit crabs do not grow their own shells. Instead, they find another creature's shell to use as a home. And, if the shell is not empty, a hermit crab might just have the occupant for lunch before moving in. As the hermit crab grows, it must find larger and larger shells. When it finds just the right one, the crab backs into its new home. If danger approaches, the animal pulls itself deep into the shell and uses its large claw to seal the entrance.

Barnacles are another kind of crustacean that often live in tide pools. They can be found cemented firmly to rocks, pieces of wood, and even the shells of other animals. Barnacle *larvae*, which look like tiny shrimp, begin life swimming freely in the water. Before long, though, they make a very tough brown glue and attach themselves headfirst to a solid surface.

Once attached, a barnacle forms a hard shell around its body and remains in that spot for its entire life. Because a

These goose barnacles use their featherlike feet to collect food. ▶

barnacle is attached headfirst, it has to eat with the six pairs of feet that stick out of the top of its shell. These feet look like feathers as they wave in the water and sweep plankton toward the animal. When danger approaches, the feet are quickly drawn inside and the "door" snaps shut.

By now, it is clear that a tide pool is teeming with life. We have seen things such as clams and crabs, squids and seaweed, sea anemones and periwinkles. Can there possibly be room for anything else in this small, watery world?

C H A P T E R   6

FROM STARS TO SQUIRTS

"I became more and more fascinated by the sea, by its mysteries. I would stare at the water's surface and wonder what lay below."

Randall Wells, dolphin expert

Let's get down on our knees and look carefully into the shimmering water of a tide pool. Do you see that little pile of sand next to the red sea anemone? That is the home of a sand castle worm. This strange worm secretes small tubes of mucus and mixes them with grains of sand. As the worm grows, it makes larger tubes and glues them to the smaller ones until it has built a sandy mansion for itself.

Lying next to the sand castle worm's house is a spiny little animal called a sea urchin. This colorful creature has

45

hundreds of tiny tube feet that help it get around. The sea urchin's mouth, which is on the bottom of its body, is full of sharp teeth for scraping algae off rocks and chopping up animal and plant matter. Some sea urchins have poison-filled spines.

The sand dollar is a close relative of the sea urchin. Both belong to a group called *echinoderms*. You can often find dried sand dollars washed up on the beach—they look like large white coins. Living sand dollars are brown and

▼ This giant red sea urchin can be found in tide pools along the coast of Oregon and Washington.

covered with short spines that feel like velvet. Mucus on these spines helps attract plankton. When predators such as sea stars approach, the sand dollar quickly burrows into the sand to hide. It takes the sand dollar 1 to 3 minutes to cover itself, but that is fast enough to escape from a slow-moving sea star.

On the undersides of a sea star's arms, or rays, are rows of little tubes with suction cups on the ends. As water is pumped in and out of the tubes, the arms move up and down, allowing the sea star to "walk" slowly across the sand. Even though a sea star does not move quickly, it is a dangerous predator. Its favorite meals include crabs, sand dollars, and clams.

When a sea star approaches, clams and other bivalves close up their shells and try to shut out the danger. The sea star uses its suction cups to grip its prey and put pressure on the shells. At the same time, it releases a chemical that makes the bivalve relax its muscles so that the shell begins to open. At that point, the sea star pushes its stomach out and over the animal. Juices in the stomach turn the helpless victim into liquid, which the sea star absorbs. After eating, the sea star's stomach returns to its normal place within the body cavity.

Even though there are few creatures that will eat an adult sea star, the toothless animals are sometimes attacked. If a sea star loses an arm it can regrow or *regenerate* that arm. Even stranger, the piece that was broken off can grow

▲ This sea star was badly injured by an attacker. In a few weeks, it will have grown brand new arms.

▼ This colorful warty sea cucumber spends its time wandering among sea urchins and other creatures. It lives in tide pools along the shores of California.

into a whole new animal! Some sea stars reproduce by tearing themselves apart, and allowing new animals to grow from the pieces.

The sea cucumber, which is related to the sea star, also displays a strange behavior. During an attack, it turns itself inside out and throws its intestines and *organs* at the enemy. While the attacker is trying to untangle itself from the mess, the sea cucumber escapes. In a few weeks, all the organs grow back. Sea cucumbers, which resemble the cucumbers you see in the grocery store, feed on decayed matter and plankton. This material is swept toward the animal's mouth by its waving tentacles.

Another curious animal is the sea squirt, which is shaped like a bag with two spouts. It can be as small as a pea or as big as a potato. Waving hairs on the sea squirt's body pull water into the top spout. Plankton is strained out

▼ This transparent sea squirt lives in gentle ocean waters along the coast of Washington.

of the water and eaten, while the wastewater is squirted out of the opening in the side. When a sea squirt is touched or picked up, water shoots out of both spouts.

Sponges have no mouths at all. They filter and absorb plankton as water flows through the holes and passages in their bodies. Even though sponges have no head, no brain, and no internal organs, they are valuable members of the tide pool. The nooks and crannies in the sponge's body provide a relatively safe home for many tiny animals.

Darting around among all the creatures in the tide pool are many tiny fish such as shannies, rocklings, and mummichogs. Shannies have no scales, so their smooth bodies can easily slip into cracks when predators appear. Mummichogs eat both plants and animals, and rocklings swim together in large groups, called schools. The entire school changes direction quickly when danger approaches.

Danger comes in many forms in a tide pool. In addition to predators, the creatures must constantly be on guard for changing weather conditions. Each species has developed unique ways to deal with the problems of survival.

CHAPTER 7

THE BATTLE TO SURVIVE

" The barnacles furl their nets and swing shut
the twin doors that exclude the drying air
and hold within the moisture of the sea."

Rachel Carson, naturalist

WAYS TO BEAT THE HEAT

Creatures in tide pools depend on water for survival. Without it, they wither and die. Sometimes the hot sun evaporates too much water or the high tide is a little lower than normal. When this happens, the animals in a tide pool must take action to save themselves.

To keep from drying out, sea anemones pull in their "petals" and turn into round blobs. This position help them stay cool and protects their tentacles from the sun. Sea

▲ As low tide approaches, the creatures in a tide pool must find ways to avoid the sun's blazing rays.

cucumbers also may curl into balls or cover themselves with mud.

Many bivalves, such as clams and mussels, snap their shells tightly closed when there is not enough water. The liquid trapped inside the shell helps keep the animal wet. Barnacles deal with dry conditions in the same way. When a fresh supply of water finally arrives, the barnacle's shell opens and its feathery legs slowly emerge to wave in the plankton-rich broth.

When the water level in a tide pool is low, some sea-weeds make a slimy mucus that coats their fronds and helps them stay moist. Other seaweeds trap moisture by collapsing. The dense mats of seaweed also provide a shady hiding place for tiny crabs, shrimp, and fish. When the tide comes in, some of the plants are lifted up again by air bladders in the fronds. The bladders act like little balloons that float in the water.

WHEN A STORM ARRIVES

Tide pool creatures can also suffer if too much water rushes into their world. During storms, huge waves crash into the pools with tremendous force. The animals must hold tight to keep from being swept out of their home. Sea urchins wedge their spiny bodies into hollows in the rocks. Snails hold tight by using their large muscular foot like a suction cup.

The sea anemone uses the disk at the bottom of its body to cling to nearby rocks. Mussels make a stringy substance to tie themselves down. They also group together in large bunches to deflect the action of the waves. Barnacles stay attached to surfaces by producing a glue so strong that the shell stays in place even after the animal dies.

Sand dollars and clams dig into the sand and stay buried until the storm has passed. Many seaweeds sway as the waves crash around them. Their graceful fronds bend and twist in the water, but do not break. Rockweeds are attached by holdfasts—root-like structures that keep them in place. The holdfasts are so strong that sometimes the plant is broken off by a wave, but the holdfast stands firm.

Life in a tide pool is seldom peaceful. Within this tiny body of water, life-and-death dramas are played out each day as the hunted try to outwit the hunters. Meanwhile, the tides and changing weather conditions threaten the survival of all the creatures, weak and strong alike.

While it is true that tide pool animals face many dangers, life can be even more difficult in the open ocean. Many more predators prowl the vast waters of the open

▼ During storms, sand dollars and other small sea animals bury themselves in the sand.

ocean. The tide pool's size offers a certain amount of protection to small marine creatures since big animals cannot fit into the small space.

Tide pools are an important ecosystem because they provide a place for tiny animals to live and grow in a relatively secure environment. The little pools also allow us a fascinating and close-up look at some of the wonders of nature. The next time you take a walk on the beach, look for a tide pool. Then, try to identify some of the animals you see. Maybe you'll be able to tell your family some interesting facts about the little creatures that live in those miniature worlds.

GLOSSARY

adapt—to adjust to a different environment.

algae—the large variety of single or multi-celled protists that do not have true leaves, stems, or roots.

ameba—a kind of one-celled, microscopic organism.

bivalve—a mollusk that has a hinged double shell, such as clams, mussels, scallops, and oysters.

cephalopod—a member of the group of mollusks that includes squids and octopuses.

chlorophyll—green matter in plants that captures energy from the sun.

ciliate—a microscopic protozoan that has hairlike outgrowths covering all or part of the body.

crustacean—a member of the group of marine animals that includes shrimp, lobsters, barnacles, and crabs. All crustaceans have hard outer shells and many "legs."

diatom—a single-celled algae that is very common in phytoplankton.

dinoflagellate—a single-celled organism that moves with two whiplike flagella.

ebb—to decline or recede.

echinoderm—a marine animal with spiny skin. Examples include sand dollars, sea stars, and sea urchins.

evaporate—to change a liquid or solid into vapor.

evolve—to develop by gradual change.

exoskelton—the hard outer shell of animals known as arthropods.

filter feeder—a marine animal that eats by filtering material out of seawater that passes through its body.

glucose—a type of sugar.

gravity—the force of attraction of one body for another.

holdfast—a root-like structure that holds some seaweeds to hard surfaces.

intertidal zone—an area of the beach found between the low and high tide marks.

larva—an animal in the first stage of life after hatching from an egg.

marine—of the sea or ocean.

matter—the substance of which an object is made.

mollusk—a member of a group of animals with soft bodies and no bones. Examples include snails, squids, clams, and oysters.

molting—to cast off the exoskeleton, outer skin, hair, feathers, or horns at certain intervals.

neap tide—a lower than average tide that occurs when the Sun and Moon are at right angles to one another.

organ—a body part made up of specialized tissues. Examples include the liver, the heart, and the lungs.

oxygen—a colorless, tasteless, and odorless gas that animals need to breathe.

photosynthesis—a process by which plants convert energy from the sun into food.

phytoplankton—microscopic marine creatures that contain chlorophyll and carry out photosynthesis; it is at the bottom of the ocean food chain.

plankton—creatures that drift with the currents and provide food for higher life forms.

predator—an animal that lives by feeding on other animals.

protist—a one-celled organism that has characteristics found in both plants and animals.

protozoan—a microscopic animal made up of a single cell or groups of similar cells.

radula—the raspy tongue of a snail used to scrape algae off of rocks or bore through shells of prey.

regenerate—to grow another.

scavenger—an animal that eats decaying organic matter.

shoal—a large group of animals, such as octopuses.

siphon—a structure shaped like a tube used for sucking and expelling water.

spring tide—a higher than normal tide that occurs when the Sun and Moon are in line with one another.

tide pool—a puddle of seawater that is isolated by the receding tide.

zooplankton—microscopic marine animals that form the basis of the ocean food chain.

RESOURCES

BOOKS

Cohat, Elizabeth. *The Seashore*. New York: Scholastic, 1995.

Rinard, Judith. *Along a Rocky Shore*. Washington, D.C.: National Geographic Society, 1990.

Rogers, Daniel. *Waves, Tides, and Currents*. New York: Bookright Press, 1991.

Sammon, Rick. *Hide and Seek Under the Sea*. Stillwater, MN: Voyageur Press, 1994.

Silver, Donald. *Seashore*. New York: Scientific American Books for Young Readers, 1993.

VIDEOS AND CD-ROMS

Ocean Planet. Discovery Communications, Inc.

The Magic School Bus Explores the Ocean. Scholastic Press.

Seashore. Boston: Dorling Kindersley Vision, 1996.

WEB SITES

Birmingham Zoo's Animal Omnibus includes information about all kinds of animals, including many of those found in tide pools.

http://www.birminghamzoo.com/ao/

Save the Tide Pools presents information in the form of a game. By the time you're done, you will have learned about the importance of saving and preserving these important miniature ecosystems and the creatures that live in them.

http://wwwbir.bham.wednet.edu/Hinshaw/tide-pool/tide.htm

The Tides provides a complete explanation of what causes tides on Earth. The site includes diagrams and a link to a site that has additional information about the Moon.

http://ispec.ucsd.edu/student-pages/our-moon/tides.html

INDEX

ABOUT THE AUTHOR

Carmen Bredeson is the author of a dozen nonfiction books for children. She has a B.S. in secondary education/English, a M.S. in instructional technology/library, and is a former high-school English teacher. Bredeson has spent many years raising funds and promoting public libraries in Texas. She is married to Larry Bredeson, a research engineer, and is the mother of a son and a daughter. The Bredesons live near Houston, Texas.